Foreword

In May 1982 the National Academies of Sciences and Engineering held a national convocation to consider the state of precollege education in mathematics and science in the United States. The objectives were to sharpen our understanding of the issues, to gain some appreciation of what might be done, and to focus public attention on the 20-year erosion of mathematics and science education.

President Reagan, in his message to the convocation, underscored the importance of the issue: "The problems today in elementary and secondary school science and mathematics education are serious—serious enough to compromise America's future ability to develop and advance our traditional industrial base to compete in international marketplaces. Failure to remain at the industrial forefront results in direct harm to our American economy and standard of living."

The convocation was not concerned primarily with preparing young people to be scientists and engineers. The more difficult and perhaps more serious problem, conferees agreed, is that we appear to be raising a generation of Americans who lack the education to participate in a world of technology—the world in which they will live and work. The costs of not preparing our children for this world will be deep and enduring, for our democratic process, for our economy, and for our security.

Changes will not be simple to make. If there is one American enterprise that is local in its design and control, it is education. Certainly there must be federal involvement, but changes will have to be at the local level, and they will be effective only if they draw on the knowledge and resources of the many other sectors of society that have a stake in how well our citizens are educated.

With this in mind, we invited people from government and industry and from education and science to participate in the convocation. More than 600 people gathered to hear 40 speakers discuss the issues and suggest ways to re-

Although there has been increasing discussion in recent months about our ability to train adequate numbers of professional engineers and scientists, we are only now beginning to see serious discussion of the far greater problem of education at the precollege level. This public awareness —and I hope public action —is long overdue.... This country was built on American respect for education.... Our challenge now is to create a resurgence of that thirst for education that typifies our nation's history Ronald Reagan

verse the current decline. The President, members of Congress, and heads of federal agencies gave their views on the role of the federal government; executive officers from business and labor discussed the contributions that the private sector can make; representatives of state and local governments and of school systems gave examples of initiatives already taken at the state and local levels; administrators and teachers in higher and lower education discussed the critical relationships between universities, community colleges, and elementary and secondary schools; and scientists and science educators provided examples and suggestions for innovations in the content of courses and methods of teaching.

Speakers and discussants from the audience focused on three topics: the nature of the problem, possible solutions, and who should be responsible for making changes. Much of the two days' discussion at the convocation was devoted to examples of activities already under way or proposed. Those examples included proposals by states for increasing the amounts of science and mathematics required in schools, mechanisms for making teaching jobs more attractive and revitalizing the morale of teachers, opportunities created by the private sector for motivating teachers and students and providing research experience, and involvement of academic science departments and scientists in improving mathematics and science education in the schools. At the national level, the National Science Board has formed the Commission on Precollege Education in Mathematics, Science and Technology; the Department of Education has created the National Commission on Excellence in Education; and the American Association for the Advancement of Science has begun efforts to increase precollege education activities by its member scientific societies.

The purpose of this report is to share the information exchanged during the convocation with a wider audience, although space restrictions make it difficult to do justice to the rich array of issues and suggestions discussed. The contributions of the speakers were key to the success of the convocation. We are grateful to them for making time in their busy schedules to participate in the deliberations. We want especially to thank Paul Hurd, who provided participants with a lucid overview (summarized in the first part of the report) on the status of science and mathematics education; Carl Sagan, who gave the inspiring special address that closed the first day's proceedings; and Gerard Piel, who had the difficult task of closing the convocation with a summary of key findings and recommendations.

It is not possible to convey in print the enthusiasm and commitment generated for us, and we believe for many other participants, by the chance to meet—sometimes for the first time—and exchange ideas with so many people concerned with giving our youth the education they need and deserve. It is our hope that this report will communicate some of the spirit as well as some of the substance of the convocation to those who will need to understand the scope of the problem, consider alternative ways of addressing it, and debate proposed changes in the many localities and sectors of our society that must take action if the state of mathematics and science education in this country is to be improved.

Frank Press
President
National Academy of Sciences

Courtland Perkins
President
National Academy of Engineering

What Is the Problem?

There are some 16,000 school districts in the United States. Each sets its own rules, often develops its own curricula, determines its own hiring and promotion practices, and fixes its own graduation requirements. Thus, even to say that a common problem exists in all U.S. schools is necessarily a simplification. Yet there is evidence that the condition of precollege science and mathematics education is not healthy. And while that evidence is necessarily aggregated, sometimes anecdotal, and often riddled with exceptions, it is persuasive and has aroused concern at the local, state, and national levels.

The evidence takes several forms. Some is quantitative: declines in Scholastic Aptitude Test (SAT) scores, declines in the number of high school graduates having taken a given number of years of science, and declines in the number of teachers certified to teach science and mathematics. Some is qualitative: in particular, the evidence for an early disenchantment with science and mathematics by children who start out being "natural scientists." This kind of qualitative finding, more difficult to evaluate than quantitative data, is perhaps more disturbing, for it indicates a profound weakness in the teaching of science and mathematics.

Test scores have been declining for several years for all subjects, although there have been some recent improvements, mainly for lower elementary grades. In two successive national assessments of knowledge in mathematics, for example, elementary school children showed a negligible learning decline, but there was a marked decline for 13-year-olds and an even greater one for 17-year-olds. In three successive nationwide assessments in science, each one showed a decline in achievement over the preceding test.

On the standardized college entrance tests, the average science and mathematics scores have been dropping steadily for 20 years. The mean score in mathematics on the well-known SAT was 502 in 1963 and 466 in 1980. Even the proportion of students scoring above 700, of a possible 800, on the SAT mathematics test declined 15 percent between 1967 and 1975. Over the same interval, students scoring below 300 on that test increased 38 percent.

But there are some bright spots: The number of students taking advanced placement exam-

We are raising a new generation of Americans that is scientifically and technologically illiterate. — Paul Hurd

Our scientific research and technological activity has been the finest in the world. We now see our technical preeminence eroding and, if we fail to act, then 10 years from now our scientific capacity will suffer similar declines. But I see a more serious threat to our nation . . . the indicators of deterioration of the quality and quantity of education in mathematics, science and technology . . . are the unmistakable harbingers of a growing chasm between a small scientific and technological elite and a citizenry uninformed, indeed ill-informed, on issues with a science component. . . . Because of the gravity of the current situation . . . the National Science Board has established the Commission on Precollege Education in Mathematics, Science and Technology to address these concerns. I place the utmost importance on the work of the new Commission. John Slaughter

inations during the 1970s more than doubled, and the mean grades in science and mathematics on those examinations increased every year. About 1.7 percent of the three million students who graduate from high school each year take the tests. These 50,000 young people provide most of the talent pool for the scientific and engineering professions. Also, high school enrollment in lower-level mathematics courses has been rising recently, particularly among girls. Enrollment of girls in engineering degree programs has increased as well.

How Much Instruction? Children are introduced to science and arithmetic in elementary school. During the 5 instructional hours in each school day, teachers devote about 45 minutes to mathematics—largely computation—and 20 minutes to science. Hence, children have an average of 1½ hours of science and 3¾ hours of arithmetic per week.

At the secondary level, most of the mathematics and science courses offered are optional. Only one-third of the nation's school districts require more than one year of mathematics and one year of science for graduation. Of the roughly three million students who graduate from high school each year, about one-third have completed three years of mathematics, and only one-fifth have completed three years of science. A scant 10 percent have taken physics, and fewer than that —8 percent—have taken calculus. Of the students in a general or a vocational curriculum—60 percent of all students—only one-fifth graduate with three years of mathematics and one-tenth with three years of science.

We, in community colleges . . . are daily confronted with a staggering number of students who cannot cope with the demands of college-level mathematics, biology, chemistry, or physics.
Ted Tilton

Furthermore, enrollments of high school students in science courses have declined in the last two decades: from 60 percent in 1960 to 48 percent in 1977. Lowered requirements for high school graduation and for college entrance and stress on grade-point averages have in part been responsibile for decreasing enrollments in these and other courses perceived to be difficult. Regardless of trends, the total enrollments in science and advanced mathematics courses are dismayingly low.

Students who continue their education try to remedy the situation in college. Remedial mathematics courses in public four-year colleges increased by 72 percent between 1975 and 1980—they now make up one-quarter of all the mathematics courses in those colleges. At two-year colleges, 42 percent of the mathematics courses are remedial. There is evidence that the greater the need for remediation, the less likely it is that a student will enter science-related career preparation.

Dislike of Science and Mathematics
When and why do students lose interest in studying mathematics and science? Apparently, students acquire a dislike for science courses early. By the end of the 3rd grade, almost one-half of all students feel they would not like to take science; by the end of the 8th grade, only one-fifth have a positive attitude toward science, and this percentage remains constant through high school. Mathematics is somewhat more popular, at least initially: nearly one-half of all students like the subject in the 3rd grade, but that too drops to less than one-fifth by the 12th grade.

Yet children like science when they learn about it on television, in museums, in planetariums, and at recreation places like marine worlds.

Why are students turned off from the formal study of science and mathematics? Several reasons have been proposed: an overemphasis on memorizing facts and technical terms; teaching abstractions that leave students puzzled as to how the science and mathematics they learn in school relate to the world; an expectation that most students will not or cannot be asked to work hard in school; a lack of respect for schools —for the intellectual effort required, for the integrity of the subject matter, for the people who teach.

Not Enough Teachers The shortage of qualified mathematics and science teachers is serious. In 1980, 28 states reported a shortage of mathematics teachers, and in 16 states the situation was listed as critical; by 1981, the number of states reporting a shortage had grown to 43. Similar teacher shortages exist for the physical sciences. During the 1970s, there was a 77 percent decline nationwide in the number of secondary school mathematics teachers being trained and a 65 percent decline in the number of secondary school science teachers being trained.

And the trained teachers who are in the classroom are leaving at a rapid rate. In 1981, almost five times more science and mathematics teachers left to take nonteaching jobs than left to retire. The replacements for these trained teachers are often inadequately prepared: of the teachers newly employed in 1981-82 to teach high school mathematics or science, 50 percent were formally unqualified and are teaching with emergency certificates. In the Pacific states, 84 percent of all high school science and mathematics teachers have only emergency certificates.

What Are Other Countries Doing? How do these conditions compare with education in some other countries? World War II left much of Europe and Japan devastated; some countries underwent civil revolution. Despite these adverse circumstances, however, such countries as the Union of Soviet Socialist Republics, East Germany, the People's Republic of China, and Japan have demonstrated that it is possible to redirect an educational system and to do so in a comparatively short time.

How much time do students in these countries spend in class? Their school year averages 240 days with minimal loss due to absences; the U.S. school year is typically scheduled for 180 days, but actually averages about 160 days because of absences. Their students attend school 8 hours a day, 5½-6 days a week; U.S. children attend school 4-5 hours a day, 5 days a week. Their school vacations are short and dispersed throughout the year to minimize interference with learning; U.S. children have a three-month summer vacation.

Specialized study for children in those countries begins in the 6th grade with separate courses in mathematics, biology, chemistry, physics, and geography. These courses last 4-6 years and are required of all students. The time spent on these subjects, based on class hours, is approximately three times that spent by even the most science-oriented students in the United

Science teaching may have become astronomy without the stars... botany without the flowers... geology without the mountains and valleys. Frank Press

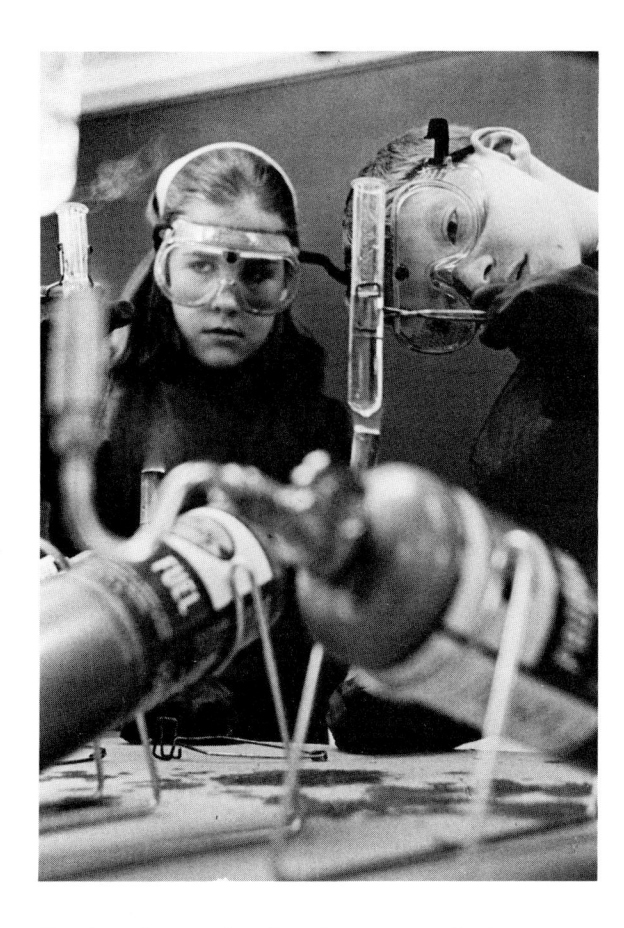

States—those who elect four years of science and mathematics in secondary school. Consequently, it is no surprise that Japan, with a population one-half that of the United States, graduates almost twice the number of engineers and the Union of Soviet Socialist Republics six times as many.

Science and mathematics teachers in the four countries are trained in specially designed programs, and each country has provisions for continuing programs of inservice education, with local colleges and universities expected to assume much of the responsibility. Members of the academies of sciences in all four countries are actively involved in keeping curriculum materials current and valid.

None of these comparisons means that educational models from abroad are appropriate for the United States, but they do tell us that other countries recognize the importance of science and mathematics education. At the very least, the programs in those countries succeed in generating both technology-oriented managers and a citizenry educated to understand scientific endeavors as well as encouraging a sizable portion of students to pursue careers in science, mathematics, and engineering. We cannot afford to do any less.

[T]he United States stands poised at the brink of an education and manpower training crisis that will make our emergence as a world scientific and technological power akin to the maiden voyage of the Titanic. . . . We won't rupture from a submerged latticework of frozen hydrogen and oxygen atoms; no, our iceberg will be a massive juggernaut of science illiteracy that will inflict a damaging blow to the ship America.

Margaret Heckler

Are There Solutions?

Raising Standards At the elementary school level, the reaction to falling test scores has been the demand for "back to basics"—teaching fundamentals. Unfortunately, the response to this demand can easily lead to an overemphasis on the rote learning of reading and arithmetic and to the neglect of science, art, music, literature, and social studies. Students cannot learn what they are not exposed to.

At the secondary school level, the trend is to increase formal requirements, but this approach runs headlong into the problem of an already crowded curriculum. To accommodate more science and mathematics, other subjects or activities have to be curtailed, or the amount of time spent in school has to be increased. Terrel Bell suggests that states consider expanding the school day and increasing summer schools for supplementary mathematics and science and also that they recommend to school systems other constructive

Let us all consider setting new maximum competency goals for our students and our citizens of enhanced math, science and computer literacy to be accomplished by the dawn of the 21st century. Terrel Bell

We must begin a broad national effort to address our growing technical human resource weaknesses....

much of the work that needs to be done will have to take place at the grass root ... parents who develop an increased awareness of the value of science and math courses for their children ... local school boards imposing tougher science/math requirements for graduation ... school administrators hiring greater numbers of highly qualified science and math teachers ... scientists and technical professionals in their own communities stressing the importance of this facet of training for our students. Don Fuqua

Only 51 percent of Oklahoma students have taken geometry, 87 percent have never taken trigonometry, and 96 percent have had no precalculus or calculus. Dave McCurdy

Universities and colleges have to join with all other components in our society in asserting that our goals in the education of our young must be excellence as well as equity.... We must decrease the chasm between secondary schools and colleges and universities. We need a sense of true partnership of all of us involved in education, viewed as one continuous and overlapping process. John Toll

ways to increase the "time on task" in mathematics and science. Are parents and students willing to accept a more rigorous curriculum, a longer school day, or a longer school year? Are educational authorities willing to impose them?

Mandatory requirements may work if they are set high enough. They will at least give young people more career choices and prepare them more adequately to be informed citizens. Several interesting suggestions for raising standards have been made. One is again from Terrel Bell: "Science [should] become part of 'basics' and basic education, and the definition of basic mathematics [should be expanded] to include problem solving and other skills beyond simple computational skills."

Some 15 years ago, states started to liberalize their requirements for graduation from high school. The norm today is a requirement of only one year of science and one year of mathematics. Some states, however, are already raising such requirements. "Leading the way down [in 1968] were California and Florida. Both states completely abandoned all state requirements for graduation from high school and many others followed the trend.... The state picture... is now at a turning point.... California and Florida... are now leading in the other direction..." (Frank Brown). California is proposing at least three years of mathematics and two years of science for high school graduation; a special commission in Florida is recommending a requirement of four years of mathematics and four years of science. At least six other states—Georgia, Maryland, Minnesota, New York, North Carolina, and Texas—are considering similar proposals.

Colleges and universities also must accept responsibility for educational standards at the precollege level. As John Toll reports: "At the University of Maryland we have been steadily raising our admission requirements in consultation with secondary schools and have given a two-year lead time each time that the standards were raised.... This year the state universities and colleges in the state have announced the gradual phasing in of additional high school course requirements of four years of English, two years of laboratory science, and three years of mathematics.... Colleges and universities can specify courses that are recommended if not required [and that]... can gradually move into the required status. Universities [also] can set requirements or recommendations for specific fields... for example, for admission to the College of Engineering, College of Business and Management, etc.... It is through such recommendations that we hope gradually to get more calculus and computer science courses into the high schools."

Attracting More Students The alternative to mandatory requirements is to make science and mathematics courses so attractive and important that high school students will be drawn to these subjects of their own accord. While this strategy tends to work best with students who are already committed, it may attract more young women and members of minority groups with potential talent and inclination for technology-based careers. Well over one-half the students in schools are girls and members of minority groups, while their representation in technical fields is disproportionately small. For example, women now make up 40 percent of the work force but represent only a little more than 2 percent of all engineers. And fewer than 2 percent of the nation's three million scientists and engineers are black.

"[I]t appears that our goal with respect to precollege science and mathematics education should have two aspects to it. All students must be guaranteed a high level of scientific and technical literacy. And the talents of those who demonstrate aptitude must be developed to their fullest regardless of whether those talents reside in a rich person, a poor person, a boy, a girl, a Black, Chicano, Native American, or a person with physical handicaps.... The United States can no longer afford to squander its talents.... I think it essential that the two aspects of this goal be pursued in tandem.... [To those] who would maintain that one aspect contradicts and precludes the other, that the same system cannot both train the extremely talented and also provide a good general education for those whose talents are closer to the norm... I would like to say that ... [we need] a symbiotic system in which the components of the system are not in contention with each other, but each part of the system is nurtured and helped to grow by every other part of the system. It is this principle that I want to propose to you as a guiding principle when tactics by which to broaden our scientific and technical talent pool are being devised" (Mervyn Dymally).

Measures that increase the opportunity for greater student participation and achievement through richer school offerings include magnet schools, programs for talented students, and summer programs. The Bronx High School of Science has a long and honorable tradition. More recently, the residential North Carolina High School of Science and Mathematics was established to serve students from all over the state, and it is becoming a local, statewide, and even a national resource. The faculty has worked with other teachers in the state on use of computers in the classroom; students devote a portion of their time to teaching others; and faculty and board members are on state and national commissions dealing with science and mathematics education.

Mathematics and science offerings for the gifted should be expanded, in cooperation with institutions of higher education. More college credit should be given for advanced courses in high school, and qualified high school students should be allowed to take university courses while still in high school.

Summer is an opportune time for flexible offerings. Courses and instruction can be provided for which time is not available in the regular schedule, thus in effect lengthening the school year. Moreover, research opportunities for able students often can be provided most readily during the summer. Research observed and done at the side of working scientists can help motivate students for further study; perhaps, more importantly, it can teach some of the processes and facts of science. Universities, government, and industry can cooperate to make research experiences both realistic and educationally meaningful.

There are many examples. "The University of Connecticut with assistance from United Technologies offers seminars for science students. The junior science and humanities symposium at Wesleyan University, supported by the research division of the U.S. Army, several engineering firms, and the university, benefits both teachers and students" (Mark Shedd). "In the Science and Engineering Apprenticeship Program of the Department of Defense [DoD], promising high school students are given the opportunity to work with the DoD laboratory researchers or university researchers under contract to DoD during the summer months. These senior scien-

We have superb resources of people and technology, of minds and material, but the help and guidance of our best minds and hearts and of our political, economic, and intellectual leadership are clearly needed to mobilize and allocate these resources to better serve the needs of our citizenry and of our nation.

Cecily Selby

We need employees —not just management and not just engineers and research people —who can think conceptually, who can reason rigorously, who are not afraid of mathematics, who are not afraid of science. Richard Heckert

tists and engineers serve as mentors. Several hundred high school students have participated in the program, and a substantial number of these young people have been women and minorities" (Caspar Weinberger).

Curricula and Teaching Even if students enrolled in more courses and devoted more time to mathematics and science, would they learn what is necessary to prepare for work and citizenship in a technology-based world? Is the content of the courses current? Is the instruction effective in conveying the content and the processes of science? Should courses be built around the traditional disciplines or should they try to interrelate science, technology, and society in new ways?

"Our methods of educating our young and the content of that education must come into the 21st century. As precollege educators we have an obligation to show our young people not only a view of the world as it is, but also to give our students a vision of what the world can become" (Floretta McKenzie).

Twenty-five years ago, a concerted effort was made to upgrade science and mathematics curricula. The effort was sparked by private foundations and, after the launching of Sputnik by the Russians, received major impetus through government support, particularly by the National Science Foundation. "[But] the world does not stand still while we improve our ability to describe it. There have been enormous advances in physical and biological sciences since the curricula in these areas were developed. There have been developments in technology that can affect education. Above all, developments in solid state physics that led to the microchip and the advent of the cheap hand calculator and cheap computer require that we rethink what we are doing, particularly in mathematics. The potential for the educational use of the computer-controlled video discs seems extraordinary" (David Robinson).

While many people agree that curricula need improving and updating, several caveats should be noted. First, a proper balance must be struck between an emphasis on content, which stresses facts, and an emphasis on process, which stresses analytical thinking. "Prior to the early 1960s, most . . . courses in science were content-oriented. The teacher would lecture, and a textbook assignment would be given. Labs consisted of exercises, and the emphasis was on facts and memorization. By the late 1960s, the approach was totally different. The student would do an investigation . . . the teacher would ask questions . . . even the textbooks were filled with questions, no answers. . . . The lecture practically disappeared. The emphasis was on thinking. Well, I haven't met many students, or adults for that matter, who like to think all the time. Both types of courses have been tried, and neither extreme could be called a success. The time has come to take the best of both . . ." (Robert Sigda).

A second caveat is that any effort to improve curricula must be sustained, and all those concerned with education—parents, teachers, school boards, and above all students—must be involved in the planning and development. Otherwise, experience has taught us, there will be little lasting change and improvement. Third, schools must find ways of investing more money in the primary teaching tools. "[E]xpenditures for textbooks have declined as a percentage of the total spent on education by 50 percent since 1965 while the educational budget has risen at an astronomical rate. Less than one cent of each educational dollar is spent for textbooks and other instructional material despite the fact that 95 percent of academic time is spent with instructional material" (Robert Bowen).

If there are to be changes in the curriculum, what should they be and how should they be accomplished? Many people believe that the content of mathematics courses should be reformulated to take account of the pervasiveness of the calculator and the computer. Children should not be taught the same mathematics in 1990 that was taught in 1960. "The use of the calculator and the computer transfers the technical problems of computation to the mechanical domain. It makes possible something we should have been doing all along but is imperative now: to teach when to multiply, not how to multiply" (Andrew Gleason).

Science curricula should be subject to continuous review and updating to bring what is being taught in line with current knowledge. In earlier attempts to accomplish this, whole new courses were created. A modular approach may be more appropriate, however. Scientific and professional societies can help by publishing pamphlets for teachers and students on new or specially interesting topics. For example, the American Associ-

Right now, as I speak, there are probably 100,000 5th grade children learning to do long-division problems. In that 100,000, you will find few who are not aware that for $10 they can buy a calculator which can do the problems better than they... faster, more accurately, than any human being can ever expect to do them.... It is an insult to children's intelligence to tell them that they should be spending their time doing this. Andrew Gleason

[A] major curriculum problem confronting science teachers is the "add-on" curriculum. These are units designed to solve many of the nation's problems. For example, if it is found that young people are eating the wrong foods, a cry is heard for nutritional units to be added on to the science curriculum. Similar calls have been made for units in the areas of drug, alcohol, and sex education, as well as conservation, energy, environmental and computer education.... The teacher is usually left to cope with the constant fracturing and disintegration of the science program. Robert Sigda

ation for the Advancement of Science plans to produce sets of notes for teachers keyed to its popular magazine *Science 82*. The National Science Teachers Association publishes separate journals for elementary and secondary school teachers; the *Journal of Chemical Education,* which used to focus entirely on the college level, now has a high school section.

Changing Needs School curricula must also take into account changes in the world of work. "Our nation is undergoing a shift in labor force needs. More and more, the members of our work force need highly technical skills. Yet, our primary and secondary schools are finding it more and more difficult to instill the basic scientific and mathematical knowledge that students must have as a prerequisite to acquiring those skills. The problem is intensified because many of those who have encountered science and mathematics education in our public schools have been so turned off by the experience that as adults they do not support measures designed to upgrade the level of education in our schools. The result is that we have a serious shortage of highly skilled professionals, a serious surplus of unemployed workers with outmoded skills, and an educational system that is increasingly inadequate to correct the imbalance" (Mervyn Dymally).

One rapidly changing field is that of clerical work. "The perception that they require little in the way of mathematics skills has been cited as one reason why women continue to go into the traditional women's fields, in spite of their poor pay and prospects. However, current technological developments, such as the increasing use of microprocessors, are having great impact on many of the traditional female occupations" (Sheila Pfafflin). "The Bureau of Labor Statistics tells us jobs requiring technological skills are among the fastest growing. For example, the demand for computer-related occupations is expected to grow from less than 700,000 in 1970 to 2.1 million in 1990. These new jobs will include many for scientists and engineers whose preparation must include highly sophisticated training.

There also will be far more opportunities for skilled technicians in computer maintenance and in repair of computer equipment. . . [and for] math technicians, tool and dye makers, and aeroastronautical engineers. The foundation for all of these skills begins with a solid background in science, math, and computer knowledge at the elementary and secondary level" (Edwin Harper).

The need to understand the use and functioning of computers is imperative for all students. Experience shows that the mere presence of microcomputers in schools, or the use outside of school of some of the most advanced microcomputer technology in the form of video games, does not necessarily lead to computer literacy. What is required is an integrated curriculum of theory and problem solving to go with the computer.

A particular need exists in vocational education to bring courses in line with current and future occupations. Control Data, for example, through the use of computer-based education in the corporation's vocational schools, has considerably reduced the time required to acquire entry-level skills for technicians in computer maintenance, operation, and programming. The company also has plans to add a robotics technician course.

Labor unions are especially concerned that today's young people acquire the skills needed for today's and tomorrow's jobs. "If we are serious about developing training programs that are more than token, we must show that the skills being taught are integral to jobs which really do exist. Otherwise, job-entry training is not likely to be considered seriously, especially by disadvantaged youth who are suspicious and hypersensitive and who are likely to assume that promises will not be honored" (Howard Samuel).

How Children Learn Until recently, little was understood of how children learn mathematics and science. Through research, a new consensus on the nature of learning is emerging.

"First, in physics and other sciences, even students who do well on textbook problems often cannot apply the laws and formulas they have

The importance of adequate skills in mathematics and its applications for today's work force cannot be overemphasized and extends well beyond scientific and technical fields. In an increasingly technological world, there are fewer and fewer jobs where such skills are not needed. Sheila Pfafflin

Children in first grade are natural scientists... then, something happens. Questions that seem interesting to a child, that are real scientific questions, get perfunctory and discouraging responses instead of the serious answers they deserve. People have been taught that they are too dumb to understand science. Carl Sagan

been drilled on to interpreting actual physical events. This observation has been made on all kinds of students, including gifted middle school children and students at some of our most prestigious universities.... Second, all students—the weaker as well as the stronger learners—come to their first exposure to science teaching with surprisingly extensive 'theories' about how the physical world works. They use these 'naive' theories to explain real world events and leave their class and textbook theories to the side when a problem different from a textbook problem is presented. What is more, studies in physics teaching have shown that students' naive theories can actually interfere with learning scientific physics. Their naive theories affect what they 'observe' [and how they interpret] class demonstrations or laboratory experiments. Third, we have learned that successful problem solving involves a substantial amount of qualitative reasoning. Good problem solvers do not rush in to apply a formula or an equation. Instead, they try to understand the situation, considering alternative representations and relationships among the variables. Only when they are satisfied that they understand the situation and all of the variables involved in it in a qualitative way do they start to apply quantification" (Lauren Resnick).

The new knowledge about how children learn science and mathematics must be translated into more effective instructional strategies. Since learners will construct meaning out of events, even in the absence of complete information, there is no way to avoid the formation of naive theories. Therefore, those theories will probably have to be confronted directly. This is a new challenge, for teaching today does not recognize that such theories exist, let alone the difficult intellectual work—analogous to what scientists themselves do—of giving up or substantially revising a theory. It may be best to start science education early, before naive theories have firmly taken hold.

Also important will be the teaching of qualitative aspects of scientific and mathematical situations. It appears that too quick an advance to formulas and routines does not help children acquire the kinds of analysis and representational skills they need. This does not mean a retreat from computational procedures in mathematics or from formulas in science, and the facts that underlie them, only that the procedures and formulas must be shown to make sense.

New Technologies Another major development in reforming instruction is the advent of computer and communications technology. These technologies are changing the way in which research in the various scientific and engineering disciplines is done, thereby requiring changes in the preparation for careers—and hence the teaching—in these fields. "The real question is whether or not this technology, properly adopted, will find appropriate usage in the classroom. Three decades ago, television was predicted to be a great educational tool. The teaching community lamented about staff reductions while administrators lauded its potential. We are hearing the same comments today about technologies in the classroom.... High technology can be used to extend the educational process—to supplement the teachers' role in new and imaginative ways. Unfortunately, this was not the case with television and, except for a very few development centers, is not now the case with the latest rage of education—the computer" (Robert Henderson). Today, the primary use of the computer is for drill. Even for this use, computers are not integrated into the total educational system in the vast majority of schools. The potential of computers for teaching problem solving and application goes largely untapped.

The continuing decline in computational costs will soon make hand-held, limited-display computers as available as calculators are today. Video discs can put the New York Public Library in every school and every classroom in the country. Continuing microminiaturization can lead to small, low-cost teaching stations that can be controlled by the learner. The revolution in communications provides opportunities to tie into regional resource centers for specialized information, lectures or demonstrations, and student testing.

All these technologies, now available, have great potential. There is a veritable grass-roots movement for their use—PTAs buying computers, after-hours computer schools and computer camps, high-technology industries establishing their own schools. The need to integrate

[C]omputer software in the real world is light years ahead of the drill-and-practice tedium of the present computer-assisted instruction. There is a crying need for computer software that meets student expectations and lets the computer become what it can instead of being flash cards on a CRT.... Remember we are talking to a generation that—one quarter after another—spent more than $4 billion on computer/video games last year. Four billion dollars to interact with CRTs. They have no fear of the technology. They are accepting it at an incredible rate, far outdistancing their teachers.   Denny Crimmins

[T]eachers, especially science teachers, need a tremendous revitalization of spirit. — Robert Sigda

Teachers will not be replaced by high technology. They are needed to promote individual treatment, interact with students at the emotional level, foster the creative process and extend knowledge acquired into real life situations.

Robert Henderson

technology products and applications into the total educational process rather than to follow the present piecemeal approach is paramount. The easy course—unfortunately all too common—is to transfer textbooks with their familiar subject matter into computer output. This serves neither the advancing state of the subject to be taught nor the pedagogic potential of the technology.

Teachers The current shortage of teachers has aggravated the poor state of science education. Explanations abound: salaries are too low to attract and retain teachers with mathematical or scientific training; the status of teachers and of education as a profession is low, and the psychic rewards for doing a difficult job are few; opportunities for relevant preservice and inservice training have become scarce.

The many concerns regarding science and mathematics teachers sort themselves into two related but separable problems: how to attract more people to mathematics and science teaching and keep them in the profession and how to improve the preparation of both new and continuing teachers.

Attracting Teachers Would paying more for mathematics and science teachers solve the first problem? Teacher organizations argue that all teachers are underpaid and that offering higher salaries only for teachers in short supply is an inappropriate way to remedy the problem. Others argue that the market mechanism is the only realistic response to current shortages and that differential salaries can be designed in such a way as to offer career advancement based on special skills and teaching contribution. A third view holds that poor morale is a worse problem than low pay.

Several school systems have developed differential salary structures. The Houston school system's "Second Mile Plan" provides extra stipends (now $800-$1,000) for qualified teachers in any area for which there is a critical staff shortage. These areas are identified each year; for 1982 they included mathematics, science, bilingual education, and special education for the handicapped. Additional stipends are paid for teachers willing to take assignments in schools with a concentration of educationally disadvantaged students, for outstanding attendance by the teacher, for obtaining academic gains by students beyond those that are predicted, and, as elsewhere, for successfully completing graduate work. With such bonuses, Houston has found that salaries for experienced teachers can be made competitive with those of local industry. Memphis is also experimenting with differential pay scales. Although the Houston "Second Mile Plan" is an approach that some teacher organizations will accept, the notion of bonus or merit pay remains controversial. Some people have suggested that each state set up a task force to study the issue of differential pay for science and mathematics teachers.

An alternative way of increasing salaries is to offer dual contracts or similar arrangements for teachers that assure them of two or three months of employment with industry during the summer. Such arrangements provide more than additional income; working in industry also puts teachers in touch with current research.

Inducements to enter the teaching profession are also needed. Special undergraduate scholarship and graduate fellowship programs could offer support to prospective teachers in mathematics, science, or other fields that suffer from teacher scarcities. Alternatively, loan programs could include forgiveness provisions to take effect after an appropriate number of years of teaching. The state legislature in Kentucky has authorized such a program of forgiveable scholarships.

Nonmonetary incentives include giving more support services to teachers, freeing them from noninstructional tasks, and recognizing outstanding performance. As noted above, some people think such incentives might be more important than added pay in retaining good teachers.

For example, teachers need better equipment, more up-to-date materials, and funds to purchase supplies. They need science and mathematics teaching resource centers—like the successful prototypes in Spencerport, New York, and Fairfax County, Virginia—that can provide low-cost instructional materials and expert resource personnel to help them try new laboratory experiments, science units, and the like.

National and state recognition programs could also be developed to honor outstanding teachers or schools. The programs would have to be large enough to make it worthwhile for teachers to participate and to be nationally visible. They should, if possible, carry not only honorific awards but prizes as well. One suggestion is that 1,000 mathematics teachers and 1,000 science teachers be selected each year for such special recognition. One type of reward could be special research opportunities or further study during the summer or a sabbatical period.

Improving Teachers' Skills Finding ways of drawing more qualified people into mathematics and science teaching and encouraging them to stay is one approach to the teacher shortage; another approach is to improve the skills of the teachers already in the system or those entering it but not certified to teach the courses they are assigned.

The Houston school district has instituted "Project Search" to identify which of its own teachers have aptitude and some college work in mathematics and science. In summer 1982, 200 teachers were selected—100 in science and 100 in mathematics—to participate in special courses offered by local universities at school sites, with the school system paying all costs. After a second summer of intensive instruction, these specially selected and trained teachers will be expected to fill some of the current vacancies in mathematics and science instruction in the Houston schools.

Summer and all-year institutes for teachers are an effective way of infusing current scientific knowledge and advances into high school courses. Institutes could be funded by the federal government, by states, locally, or by industry. For example, Sun Oil Company sponsors seminars

I think it is shortsighted to expect talented teachers of the caliber that we're demanding in our school systems to be willing to live lives of genteel poverty just because that's what people expected school teachers to do when I went to school. That won't happen anymore.

Patricia Shell

Right now, our best teachers in science and mathematics are being motivated to seek careers in industry. It takes little insight to see that, ultimately, this trend will be destructive both to industry and to education. The disaster that awaits us might yet be avoided if educational institutions and industrial institutions resolve to deal with each other as partners rather than as rivals for personnel. Mervyn Dymally

It is my claim that the university must share responsibility for the input to their doors, an input they must have.... I do not think it wise, safe, or workable in the long run to neglect the coupling between the university and the early schools. Setting requirements and exams is merely a necessary condition; sufficiency means concern, knowledge, and genuine helpful participation. Philip Morrison

for science teachers during the annual conventions of the National Science Teachers Association. The General Electric Company runs an inservice program for school staff to provide educators in G.E. communities with knowledge about industrial careers and the impact of technology on the workplace. Each program is tailored to the needs and the resources of the locality, and graduate credit is given through cooperating local colleges. In Arizona, utility companies sponsor the development of science units and inservice programs for teachers based on these new materials. As an alternative to completely local sponsorship, the federal government might be responsible for the costs of updating the content of teacher-training courses and local governments or industry for the costs of the institutes and curriculum materials themselves.

Informal training opportunities are also needed. Teachers should be encouraged to participate in conventions, conferences, workshops, and teacher exchanges between schools. "The enthusiasm alone which teachers bring back to the classroom following such activities more than makes up for the break in continuity and learning while they are away. Indeed, these teachers usually bring back a briefcase full of new ideas designed to stimulate their students. Participation in all these activities must not only be encouraged, they must also be funded. Presently, most science teachers pay their own way to participate in these activities, be it a college course or attendance at a convention. As a result, the number of science teachers participating diminishes each year" (Robert Sigda).

Work in industry, government, or university laboratories during the summer or during part of the school year on a released-time basis should be made available to teachers. Besides augmenting their income, such opportunities can keep good teachers in teaching by renewing their knowledge and enthusiasm. And not only will they bring more up-to-date instruction to their classrooms, but they will also be able to provide more realistic advice to their students on career possibilities.

"Extra" Teachers Sources outside the teaching profession can be tapped to contribute to schools. Using university students and scientists and engineers from industry would help alleviate the teacher shortage, bring working scientists or engineers into the classroom, and strengthen ties between schools and industry and institutions of higher education. Houston has established a "Teaching Partners" program in which a full-time science teaching position is shared by two people. Some individuals who have left teaching for jobs in industry return and teach half-time with the approval of their private employer. In Los Angeles, Atlantic Richfield Company sends 100 of its employees into four inner-city schools to teach mathematics and special courses and to motivate students. In some cases, retired scientists from industry or government also may be interested in part-time teaching.

Since most such recruits to teaching lack traditional teaching credentials, their work in a school system may create controversy. States should consider competency-based (rather than credit-hour-based) teacher certification; an alternative is to create a separate category, say, adjunct teacher, comparable to an adjunct professor in a university, for scientists who want to teach in schools but lack traditional certification.

Undergraduate and graduate students represent a tremendous potential resource. One idea is that work with schools be allocated 5-10 percent of the research resources of a university through the contribution of student and faculty time spent in the secondary schools in such activities as working with teachers on curricula and experiments.

Another proposal is that science-trained college graduates work in elementary school classrooms in exchange for fellowships for further study. Graduates with more advanced degrees would serve as senior science resources. "Bachelors graduates would get a... fellowship of $9,000/year tax free for a minimum of two years and a maximum of three years. Masters graduates would get $12,000, and doctoral graduates $15,000. In addition, all would get

$3,000/year credit toward past or future school loans. The total cost of putting 77,000 science-trained people in all our elementary schools would be $982 million/year.... [T]he cost [should] be split—the federal government's share would be $482 million—leaving about $25,000 per district to be raised through cooperative school district and business/industry efforts. Total cost to school districts—less than one-half percent increase of the average $3,000,000 district budget" (Richard Ruopp). Such a yearly investment would ensure that every elementary school in the country had a science-trained college graduate in contact with students from kindergarten to 6th grade.

Outside the Classroom In addition to improving what goes on in school, there is a need to introduce children to scientific concepts early, to provide opportunities for out-of-school experiences with research and with the application of formally acquired knowledge, to illustrate the role of science and technology in the working world, and to generate pride and enthusiasm for the learning of science and mathematics.

Video computer games appeal to children and adolescents. "There must be a way to devise a computer game which will be as attractive as any on the market now that will illustrate scientific concepts.... Scientists simply have not engaged these very powerful educational tools" (Carl Sagan). Space vehicles in video games could be programmed to operate in a Newtonian gravitational field and provide a sense of the inverse square law, of a parabola, of a hyperbola, of an ellipse. Devising a game program based on the world of the very small, in which the laws of nature are those of quantum mechanics, would provide a glimpse of a counterintuitive world, giving—as Lewis Thomas recently urged—mysteries before the facts, so that children learn "there are some things going on in the universe that are beyond comprehension."

Specially designed television and film materials that illustrate the challenge and joy of doing science appear to work well with children in upper elementary school. For example, the nine films in the "Search for Solutions" series let viewers work alongside some 120 scientists in pursuit of a solution. Phillips Petroleum underwrote the production and has been distributing the films free, together with a teaching guide. The series has had more than 2 million showings, and about 54 million people have seen it. It also was a success as a commercial movie feature in New York. A similar series in mathematics, also supported by Phillips, is being made. But much more is needed. In some school districts, "Search" has turned out to be their only science

What we need are bright, energetic, dedicated young people, trained in mathematics... science... or technology, mixing it up with 6- to 13-year-old kids in the classrooms, behind telescopes, by ponds and tidal pools, and at the microcomputer. Richard Ruopp

[S]cience museums present information.... They excite people.... Museums let people learn in their own way, at their own pace, at their own schedule. Joel Bloom

program because fiscal stringencies have forced media libraries to close and have pared science budgets virtually to zero.

Science and technology museums, zoos, planetariums, marine exhibits, and similar places have an important role in stimulating interest in scientific processes and phenomena. These institutions provide experiences that teach both understanding and enthusiasm for science.

For older students a better understanding of career opportunities must be developed, especially in the 7th and 8th grades. It is at this stage that young people often cut off possibilities for entering science- and mathematics-related fields because they fail to enroll in the prerequisite high school courses. The efforts of school counselors and teachers can be much strengthened by attractive audiovisual materials dealing with real jobs and what it takes to get them.

"[T]he Electronic Security Command of the Department of Defense has developed a 30-minute audiovisual briefing designed to encourage 7th and 8th graders to take more science and math courses.... It uses a combination of words, music, and slides to illustrate how modern technology has permeated virtually every part of their lives, from electronic games to nonstick kitchenware, from hand-held calculators to zero calorie foods.... It also takes a look at future technology and how students may directly contribute to constructing space stations, building robots, or finding new medical cures" (Caspar Weinberger).

The opportunity to visit worksites—industrial plants, government or university laboratories—and, preferably, to gain work experience is a very effective way of having young people learn about career opportunities. Work-study programs can be offered on a part-time basis during the school year and full-time during the summer. In the past, such opportunities were made available to small groups of science-talented students through government funding, and industry also has provided work experience, often to young people needing an income. Such programs need to be expanded to reach many more students.

[O]ur youngsters need to be exposed as early as possible to science, math, and technology. Unless we begin now to motivate and equip them to pursue scientific and technological careers, the shortages will persist. The national lesson Sputnik taught us in 1957 should still serve us. Caspar Weinberger

Recognizing Achievement Not only meritorious teachers should be honored. Recognition of outstanding students would help generate more interest for all students and implicitly recognize their teachers and schools as well. Again, such opportunities exist at the very top—for example, the Westinghouse Talent Search. This is far removed, however, from the aspirations of most students. Science fairs, prizes, even recognition for completing successfully a rigorous mathematics- and science-based curriculum are all means to highlight achievement.

The University of Maryland runs statewide examinations in mathematics on which cash prizes and scholarships are based. Teachers of the winning students also get special public recognition. In Connecticut, some 50 businesses and other organizations support the state's science fair. "[T]he National Science Teachers Association, under contract with the National Aeronautics and Space Administration, has been and is conducting national competitions in the nation's secondary schools. The students design experiments for the space shuttle, and the winner's design is developed and flown aboard the shuttle" (Robert Sigda).

One suggestion that has been made is for a science and technology fitness program analogous to the President's Physical Fitness Program, with presidential endorsement and awards for having taken four years of science and four years of mathematics.

Meeting the Challenge: Whose Responsibility?

Just as the problem of precollege science and mathematics education is not one problem, there is no one solution. As demonstrated by the many examples above, however, there are a variety of effective responses: experiments tried by one locale that are transferable to another, curriculum changes that are promising and now demand validation, innovative approaches to teaching that exploit new technologies. The issue is how to meet the challenge at a time of intensifying economic constraints, locally and nationally.

The Public Effective action will come about only when the general public becomes sufficiently concerned about the existing crisis—a "scandal," Carl Sagan calls it—to demand improvement. For people to care about educating their children in mathematics and science, they must appreciate the role of science and technology in modern society, in the world of work, in international markets, in people's health and longevity. Therefore, initiatives to improve science and mathematics education in schools must be coupled with programs to further public understanding of science and technology.

There have been some excellent and very popular science programs on television, such as "NOVA," "Life on Earth," and "The Body in Question." The "Cosmos" series has been seen by something like 20 percent of all Americans. Attendance at science museums keeps increasing. The proliferation and success of science magazines for popular audiences has been a recent publishing phenomenon. As Carl Sagan has observed, the popularity of science programs, magazines, and museums testifies to an unsatisfied hunger for science, an intuitive understanding of its power and joy.

Regional and local conferences should be held or task forces organized to try to build consensus on the need to improve science and mathematics education and to consider ways to do so. Task forces should include representatives of education, business, labor, government, and private organizations. Community and state leaders are the key to improvement: exhortations from Washington and federal mandates are not enough.

> *[I]mproving science and math achievement among our young people requires a joining together of efforts by educators, parents, the private sector, and all levels of government. . . . [We need] a continuing national discussion on the importance of high-quality math and science instruction in our precollege educational facilities.*   Edwin Harper

> *[P]eople . . . understand very well what science is about. It is a tool for managing the future. It is extremely powerful. It is a way of participating in the technological civilization we inhabit.*
> Carl Sagan

Many Roles, Many Actors Given national agreement that action is needed, even given some agreement as to what actions would be most effective—Who ought to do what? Where will funds come from? In light of the kaleidoscopic structure of the country's educational system, responses are not the province of any one set of actors.

Arguably, new partnerships are needed between education and industry, between scientists and educators, between educational institutions at different levels, between state and local government and business, and between the federal government and all the other sectors. Examples of such partnerships abound in rich variation. All recognize the local responsibility for education. Beyond that, points of view differ on who needs to do what.

Government There is an economic argument for public investment in education in general and for mathematics and science education in particular, especially in light of the country's productivity decline. "Anyone seeking ways to achieve a resumption of productivity advance might be expected to consider all sources of past growth. One might expect government, when considering its role in an attack on the problem, to give special attention to output determinants for which it bears primary responsibility. If government does so, it will inevitably focus on education, which is at once an area of primary government responsibility, the second largest source of growth, and an influence... on the largest source, the advance in knowledge" (Edward Denison). In considering the role of "government," the presence of different levels of government complicates the situation.

I'm convinced that we won't see major improvements in schools until parents, teachers, local administrators and the students themselves understand the problem and demand higher levels of performance.

Stories of kids graduating from high school without being able to multiply or divide make people mad. National study commissions are less important than grass-roots drives for improvements. Richard Heckert

[A]s we move along the scale from local to federal government, we encounter the deep tradition that precollege education is a local concern that is very closely coupled to parents and to home. Finding an acceptable role for all levels of government within those traditions and within our pluralistic educational modes is urgent considering the situation.... Edward David

The American economy is in trouble. And this means that America is in trouble.... [M]y concern... is with America's long-term growth.
Edward Denison

23

We need a national policy on technical, scientific and engineering manpower and education....

I hope to spark the kind of debate that will lead to a consensus on precisely what [the] federal role ought to be. In my judgment, this is one issue on which there can be no delay or complacency. Action is needed now if we are to halt the erosion in the store of human and educational resources that have made America the great nation it is.

John Glenn

A key federal responsibility is the support of a strong, basic research program.   Harrison Schmitt

"The federal government is the primary source of support for basic research, which provides the basis for future inventions and discoveries.... Other federal responsibilities are more directly related to precollege science and mathematics education. The role which the National Science Foundation played in the curriculum development in the early 1960s was critical to the enormously productive technological effort which the nation made, partly as a response to the challenge of Sputnik.... Improving the knowledge and skills of those who teach science and mathematics in the primary and secondary schools is a responsibility shared by the federal government and the states. The progress of science and technology is now so rapid that school systems are hard pressed to keep pace with the advances. The federal government is in a much better position in this regard to develop information on behalf of the states' inservice training efforts.... Support of young promising faculty and inservice training for precollege science and mathematics teachers are good examples of where well-designed investments should have enormous influence. Further, such investments should be shared by the private sector, and there is increasing evidence it is doing so and will do more" (Harrison Schmitt).

At the other end of the scale, local school boards are ultimately responsible for what goes on in schools. The turnover rate among the 95,000 school board members that make up the 16,000 local school boards is 25 percent each year. "[There are] a lot of new people on board each year, learning for the first time some of the problems [they] face.... [T]here is a tremendous frustration level among school board members. Most school board members leave, or choose not to run for board office, or even not to run for reelection because of frustration.... [T]heir frustration is rooted in not having an impact on the curriculum" (Alan Shark).

It has been suggested that school boards and other local leaders need tools that allow self-assessment of the status of science and mathematics education in a given school system. Once they have identified weaknesses, they also need information on responses that have been successful elsewhere and are appropriate for the particular system and its weaknesses.

Public/Private Partnerships "Society has a fundamental stake in the quality of education in this country and solutions must be broadly based if they are to be effective and long lasting. Our goal should be to develop new kinds of bonds between the private sector, which is the ultimate user of human capital our schools develop, and the schools, which are the sources of that capital" (Ronald Reagan).

What can other sectors do to help local authorities improve the schools? Local task forces could initiate activities between local business and school districts: loaning technical employees to schools, making part-time work and research opportunities available for teachers and students, keeping teachers abreast of recent advances, supporting recognition programs that will reward excellent teachers and students. Business and industry employ 60 percent of all scientists and engineers, nearly 1.9 million people. If even one-tenth of this resource became active in precollege education, every school in the country would have access to two or three scientists or engineers.

There are a number of examples of cooperative public/private ventures sponsored by industrial organizations. CIBA-GEIGY Corporation sends scientists to visit schools and invites science classes to the company. They also run summer internships for students and a national awards program for teachers. Eastman Kodak sponsors special programs for young people, including career exploration and summer employment for talented high school students. In addition, they are part of a Rochester consortium of 51 industries that emphasizes curriculum development, teacher training, and work with parents to encourage science and mathematics learning by minority children from the 6th through the 12th grade. In Boston, industry is heavily involved in the school system's new occupational resources center. Curricula for technical occupations are largely written by area employers; Digital Equipment Corporation is contributing equipment and one of its full-time employees to consult at the center; Apple Computers also has donated equipment.

Parallel efforts in partnerships between industry and education must go forward at the state level, as they already are in a number of states. The National Governors' Association reports 88

[S]olutions must involve fundamental changes in the relationship between education and business. Education is everyone's business and everyone's responsibility. And hence, there must be much closer working relationships among corporations, schools, and other sectors . . . . William Norris

Wellspring...should not be viewed as a traditional economic development venture. Rather, it is a collaborative effort by leaders in labor, business, education, and government to find common ground —and then sanctify that meeting ground by pledging that old and natural battles will not be fought on it. Wellspring is a demilitarized zone born of the recognition that extraordinary measures are required if we are to maintain and improve our technological and economic strength in light of severe competition. Albert Quie

state initiatives under way to increase technological innovation and productivity. For example, Minnesota has established Minnesota Wellspring to further the state's economy and leadership in technology; education is one of five areas for action. Virginia has just announced a comprehensive program to develop bridges between the state's education community and high-technology industry; California and Florida both have commissions with wide representation to examine precollege science and mathematics education.

A more far-reaching suggestion for involvement of the private sector has been made by William Norris: "The missing ingredient is nationwide broad-based partnerships in which business addresses improvements in education as profitable business opportunities in cooperation with other sectors. In order to realize the full advantage of the use of advanced technology in the educational process, the management of schools themselves, or schools within schools, should be contracted out to business, which has the expertise to use advanced technology efficiently. Such an arrangement need in no way negate the responsibility of teachers to diagnose students' needs and to select curriculum material. In addition to the use of technology in teaching, [business] can assist with record keeping, testing, and other chores, thus permitting teachers to be free to do what they do best: teach." Control Data Corporation has invested

[T]he communities engaged here in the conference —representatives and spokesmen from industry, from the scientific community, the education community, the local, state and federal government —all agree that each has responsibilities, and each has testified to the action and commitment that they have to this task. Gerard Piel

more than three-fourths of one billion dollars in educational programs for PLATO, an innovative computer-based instructional system originally developed at the University of Illinois with support from the federal government—a good example of partnerships.

Colleges and Universities Colleges and universities traditionally have affected curricula at lower levels by setting college-entry requirements; they also train teachers. A neglected responsibility is providing relevant curricula for college students planning to become teachers. The academic preparation of mathematics and science teachers typically does not include the range of courses they need to do a good job of teaching science and mathematics in secondary school.

Science teachers, like engineers, are in an applied field. "Engineers select information from the sciences and mathematics to solve problems. Teachers select information from these same fields to interpret science to nonspecialists. However, the education of a science teacher, an interpreter of science, is that of a researcher cut short at the end of four years" (Paul Hurd). University science departments and schools of education must recognize their responsibility in reforming teacher education. "[E]very major university should have in each math or basic science department at least one outstanding faculty member whose major responsibility it is to work with the schools in the teaching of the discipline.

At the University of Maryland most of these faculty members have a joint appointment between the basic science department and the science teaching center of our College of Education" (John Toll).

Another important responsibility of colleges and universities is to provide additional educational opportunities in mathematics and science—enrichment and advancement for talented high school students, remediation for those who come to higher education not adequately prepared. Community colleges, in particular, have a special role to play in providing education for technology-related careers.

Scientists "To assert the priority of scientific literacy is not to attempt to impose upon American education the aims of yet another single-purpose pressure group. On the contrary, it is a call on American society to redeem its promise to its children: that is, to fulfill their right to the best education society can provide.... The liberating objective of scientific literacy cannot be accomplished by a one-time effort.... What is required is the permanent, sustained, and increasing commitment of the American scientific community..." *(The State of School Science, 1979).*

The first wave of science education reform was initiated by scientists after World War II when they became aware of the failures of high school science instruction. Later, Sputnik rallied government and the public to the cause of reform. A similar commitment is needed now, and scientists need to lead the way. Scientists should make precollege education part of their professional responsibilities. They should spark and contribute to updated curricula, teacher education, student motivation and research opportunities, and public understanding of science. To ensure the health of science and mathematics education in the schools, they ought to participate in the governance and improvement of local public schools. Scientists in the various disciplines taught in school and researchers in the cognitive sciences should continue research on how science and mathematics are learned and should help in applying the findings in the schools.

Scientific societies furnish an effective mechanism for involving the scientific community in precollege education. They can "provide leadership in rethinking the objectives and goals of precollege education... enhance the prestige of precollege science teachers within the scientific and technological community... ensure the integrity of science... [and] promote an understanding amongst those who teach of the relation of science and technology to society" (Anna Harrison).

In the final analysis, it will be a wide range of institutions, with the support of the American people, which will bring about the progress all of us seek in improving precollege mathematics and science education. The answer lies in the imaginative initiatives undertaken to bring together all levels of the private and public sectors to achieve the goal we all share. Ronald Reagan

Speakers

Opening Statements

Frank Press
President, National Academy of Sciences

Floretta D. McKenzie
Superintendent of Public Schools, District of Columbia

The State of Precollege Education in Mathematics and Science

Paul DeHart Hurd
Professor Emeritus, Stanford University

Perspectives on the Problem—An Overview

Edwin L. Harper
Assistant to the President for Policy Development, The White House

William C. Norris
Chairman of the Board, Control Data Corporation

Albert H. Quie
Governor of Minnesota

John B. Slaughter
Director, National Science Foundation

Terrel H. Bell
Secretary of Education

Caspar W. Weinberger
Secretary of Defense

Congressional Viewpoints

Harrison H. Schmitt
United States Senator, New Mexico

Don Fuqua
United States Representative, Florida

Margaret M. Heckler
United States Representative, Massachusetts

Dave McCurdy
United States Representative, Oklahoma

The Role of Government: Federal, State and Local

Edward E. David, Jr.
President, Exxon Research and Engineering Company

B. Frank Brown
Chairman, Governor's Commission on Secondary Schools for the State of Florida

Patricia M. Shell
Superintendent for Instruction, Houston Independent School District

Mark R. Shedd
Commissioner of Education, Connecticut

Edward F. Denison
Senior Economist, Brookings Institution

Cecily Cannan Selby
Co-Chairman, National Science Board Commission on Precollege Education in Mathematics, Science and Technology

John H. Glenn, Jr.
United States Senator, Ohio

The Role of the Private Sector: Business and Industry, the Media, Foundations

Richard Heckert
Vice Chairman, E.I. du Pont de Nemours & Co.

Howard Samuel
President, Industrial Union Department, AFL-CIO

David Z. Robinson
Executive Vice President, Carnegie Corporation of New York

James C. (Denny) Crimmins
Producer, "Search for Solutions"

Robert C. Bowen
Vice President, Marketing, McGraw-Hill

Joel N. Bloom
Director, Franklin Institute Science Museum and Planetarium

Special Address

Carl Sagan
David Duncan Professor of Astronomy and Space Sciences and Director, Laboratory for Planetary Studies, Cornell University

Expanding the Scientific and Engineering Resource Base

Mervyn M. Dymally
United States Representative, California

The Role of the Scientific Community

Anna J. Harrison
President-Elect, American Association for the Advancement of Science

Robert P. Henderson
Chairman of the Board and Chief Executive Officer, ITEK Corporation

Sheila M. Pfafflin
District Manager, American Telephone and Telegraph Company

Philip Morrison
Professor of Physics, Massachusetts Institute of Technology

Andrew M. Gleason
Professor of Mathematics, Harvard University

Lauren B. Resnick
Professor of Psychology and Education, Co-Director, Learning Research and Development Center, University of Pittsburgh

The Role of the Education Community

Robert B. Sigda
President-Elect, National Science Teachers Association

Alan R. Shark
Director of Research and Board Development, National School Boards Association

Richard Ruopp
President, Bank Street College of Education

Ted Tilton
Provost, The College of DuPage, Illinois

John Toll
President, University of Maryland

Closing Remarks

Gerard Piel
Publisher, *Scientific American*